John and Mary Gribbin

MENDEL
(1822–1884)
in 90 minutes

Constable · London

First published in Great Britain 1997
by Constable and Company Limited
3 The Lanchesters, 162 Fulham Palace Road
London W6 9ER

ISBN 0 09 477120 0
Set in Linotype Sabon by
Rowland Phototypesetting Ltd,
Bury St Edmunds, Suffolk
Printed in Great Britain by
St Edmundsbury Press Ltd,
Bury St Edmunds, Suffolk

A CIP catalogue record of this book
is available from the British Library

Contents

Mendel in context

If you know anything at all about Gregor Mendel, the chances are that you know that he was an obscure Moravian monk with an interest in gardening, who dabbled in pea-breeding experiments in the middle of the nineteenth century, and happened to notice the way that certain characteristics were passed on from one generation to the next. You may have heard that there is some doubt about the accuracy of his results, and that he may have fiddled the data to get such good statistics. You probably also know that nobody had heard of his work until it was rediscovered in 1900, some 16 years after Mendel died. And you'd be wrong on every count.

Far from being obscure, Mendel rose to become abbot of his monastery, was politically active at a time of great change in the Austro-Hungarian Empire, had a long and public battle with the government over the taxation of religious orders, and travelled widely, visiting both Rome and London. Far from being a gardener with a dilettante's interest in science, Mendel was a trained scientist with a thorough understanding of physics and statistics, who applied the methods of physics to his plant-breeding experiments, turning such studies from an art into a science. He didn't stumble on peas as suitable specimens for study, but chose them for his detailed work after extensive observations of different kinds of plants (and some work with mice), precisely because he knew that in peas the characteristics he wanted to study were passed on 'cleanly' from one generation to the next.

Although the accuracy of his results has been doubted (the claim being that they were too good to be true), it is now clear that although Mendel's modern critics were able statisticians, they didn't understand plant breeding, and misrepresented him. They had forgotten to allow for the fact that about one seed in ten fails to germinate, and this natural wastage neatly explains the apparent discrepancies in Mendel's published results.

As for being unknown, Mendel corresponded with one of the leading botanists of his day, Karl Wilhelm von Nägeli, who was impressed, and included some of Mendel's conclusions in his own books (although not, alas, with sufficient credit to Mendel). The work was discussed by other biologists, and even merited a mention in the *Encyclopaedia Britannica* well before the end of the nineteenth century. Mendel himself was well up to date with the latest developments in evolutionary biology, reading all of Charles Darwin's work as it was translated into German; but his scientific activity almost stopped once he was elected abbot in 1868.

One stumbling block did, perhaps, hold back the development of evolutionary theory for thirty years. Darwin himself, the person best placed to appreciate the value of Mendel's work, never read Mendel's publications, and missed an opportunity to reinforce his theory of evolution by natural selection with crucial supporting evidence, at a time when it was being widely attacked. It is, of course, the combination of Darwin's theory of natural selection and Mendel's discovery of the rules of inheritance and genetics that makes the modern theory of evolution so successful as a description of nature.

By the time Mendel began his major work, in the mid-1850s, the idea of evolution was very much in the air. On the theoretical side, it was widely thought that offspring inherited a mixture of characteristics from each of their parents, somehow blending the characteristics of their parents together. It was also widely thought that the influence of the environment on the characteristics of animals and plants as they grew would be passed on to succeeding generations. If, for example, an animal was brought up in a cold climate, and responded by eating lots of food and laying on a thick layer of fat, its offspring were expected to be born with a ready-made, extra-thick fat layer. In practice, farming had become increasingly scientific, and there were sound commercial reasons to be interested in ways of improving the stock of both plants and animals by selective breeding for desirable characteristics – but it all depended on empirical rules of thumb.

These developments were among the main sources of inspiration for Darwin himself, who became a skilful breeder of fancy pigeons. Still, when the *Origin of Species* was published in 1859, it included his admission that 'the laws governing inheritance are for the most part unknown'. At that time, Gregor Mendel was almost exactly in the middle of his classic series of experiments that would lay bare those laws.

Life and work

The man we know as Gregor Mendel was born in Heinzendorf, in Moravia (then part of the Austrian Empire) on 22 July 1822.

Almost everything in that sentence needs qualifying. The baby was actually christened Johann, and the young Mendel took the name Gregor only when he became a novice monk. Heinzendorf is now known as Hyncice, and is in the Czech Republic. And although Mendel always celebrated his birthday on 22 July, his baptismal certificate records the date of birth as 20 July. But there is less confusion about the Mendel family background, and the boy's upbringing.

The word 'peasant' today conjures up an image of the lowest kind of land worker, and is sometimes used in a pejorative sense. In fact, being a peasant in Moravia in the nineteenth century (which is what Mendel's father, Anton, was) was a good step up the ladder from such a lowly occupation, the crucial point being that a peasant actually held land and farmed it for himself. The conditions under which Anton Mendel and his like farmed their modest plots were pretty grim, even so. Apart from anything else, they each had to spend three days a week working for their landlord.

This, like the entire Austrian Empire, was a hangover from the feudal age; Anton Mendel himself had been affected by the great changes occurring in Europe in the nineteenth century. Born in 1789, he had served as a soldier for eight years, old enough to have played a part in the Napoleonic Wars, travelling widely enough to learn that there was more to life than being a peasant farmer. In 1818, Anton married Rosine Schwirtlich, the daughter of a gardener in Heinzendorf (we shall stick with the names of places as they were in Mendel's day, rather than updating everything). Their first two children, both girls, died in infancy. The first to survive, a girl christened Veronika, was born in 1820. Johann came along in 1822, and another daughter, Theresia, in 1829.

Moravia is one of three geographical units which together essentially made up the former Czechoslovakia. It lies between Bohemia to the west and Slovakia to the east, and Bohemia and Moravia together pretty much make up the Czech Republic. Heinzendorf is in the north of Moravia, and as the name indicates this was a largely German-speaking area in the nineteenth century. The Mendels spoke German as their first language; Johann regarded himself as German, and wrote in German, although he spoke Czech as his second tongue.

Although busy with his work for the landlord and his own farm, Anton Mendel had developed an interest in growing fruit, and improved his small orchard with grafts supplied by the parish priest, Fr Schreiber; helping his father with this work was probably young Johann's introduction to horticulture.

Schreiber also taught at the village school. By the standards of the Austrian Empire in the 1820s, this was a relatively enlightened institution. At the behest of the local lady of the manor, Countess Truchsess-Zeil, the children were not only taught to read and write, but were given an introduction to natural history and science.

Mendel lapped all this up, and was an outstanding pupil. Encouraged by Schreiber, his parents allowed him to go, at the age of 11, to a more advanced school in Leipnik, about

20 kilometres away. Although Johann was their only son, and Anton wanted to pass his peasant holding on to him, his eyes had been opened by his experiences in the Napoleonic Wars, and he did everything within his limited power to secure a better future for the boy. Johann was again an outstanding student in Leipnik, and in 1834 he moved on to the *Gymnasium* (High School) in Troppau, 50 kilometres from his home.

This resulted in real hardship. His parents were unable to support Johann fully, and he had to work for part of his keep, but he continued to do well academically. Then, in 1838, when Johann was 16, disaster struck. Anton was severely injured in an accident, and had to give up farm work. Johann had to provide for himself almost entirely (how, we are not sure, but probably as a tutor to younger children) and became ill, probably through a combination of overwork and poor nutrition. But he completed his *Gymnasium* studies in 1840, and enrolled in a two-year course at the Philosophical Institute in Olmütz, designed to lead to university entrance.

Again, Johann had no problems with the

courses at the Institute, where he gained a thorough grounding in, among other things, mathematics, physics and statistics. But again he suffered from ill-health (presumably for the same combination of reasons as before), and by the end of the course the financial situation was desperate.

With Johann bent on an academic career, the injured Anton Mendel had made over the peasant holding to his son-in-law Alois Sturm, who had married Veronika. The terms of the sale (dated 7 August 1841) show that the family had already got a pretty good idea which way the wind was blowing. They specifically included a requirement for Alois to pay 100 florins to Johann if and when Johann should enter the priesthood. The contract also provided for a dowry for Theresia when she married. However, Johann managed to complete his studies in Olmütz only because Theresia, almost immediately the contract was signed, voluntarily gave up part of her dowry to pay for his last year there. At the end of that year, the family's resources were utterly exhausted, and the priesthood offered the only hope.

Now, this wasn't quite such a drastic step

as it may appear to modern eyes. What was probably in Mendel's mind initially was that as a priest he might hope to become a schoolteacher, like Fr Schreiber, and carry out some scientific studies in his spare time. After all, Mendel was a good Catholic, and would have no moral problems with the religious aspects of his new profession. But things turned out better than he could have dreamed. Mendel, who now regarded himself as a physicist, sought help and advice from his physics professor, Friedrich Franz. Franz had spent many years teaching at the Philosophical Institute in Brünn (now Brno) in southern Moravia, and while there he had lived with and got to know the monks of the Augustinian monastery. But this was not your average monastery.

The monastery of St Thomas in Brünn was one of the leading centres of intellectual activity in Moravia, and influential on the spiritual side of life. The abbot, Cyrill Franz Napp, had taken over as head of the community in 1824, when he was only 42. This was normal practice at the time. Because a heavy tax was levied on a monastery when its prelate died, the monks would deliberately elect a young man to the position so as to postpone the next day of reckoning.

Napp took over a run-down institution in some financial difficulties, turned it into a flourishing centre, and encouraged the members of the community to develop artistic or scientific interests. The community included a botanist, an astronomer, a philosopher and a composer, all well regarded by their peers. And under an Imperial decree of 1802, the monks were required to teach Bible studies, philosophy and mathematics at the Philosophical Institute in Brünn – ample justification for their activities. Furthermore, the monastery came under the direct rule of the Augustinians in Rome, exempting it from the day-to-day interference that might otherwise have come from the Church authorities in Austria.

Even so, it is amazing what Abbot Napp got away with, virtually creating a small private university around the dozen or so monks of his community. In 1843, he was also on the lookout to add to his small band of intellectuals, and asked Professor Franz if he could recommend any young men from among his students who might like to join the order.

There was only one that Franz felt able to recommend, 'a young man of very solid character' whom Franz described as being 'in my own branch [physics] almost the best'. On 9 October, Mendel was formally admitted to the order as a novice, taking the name Gregor, and being known thereafter as Gregor Johann Mendel.

Several years later, in an autobiographical essay, Mendel wrote that he had 'felt himself compelled to enter a station in life which would free him from the bitter struggle for existence'. He also wrote that this step 'brought a radical change in his circumstances'. He would certainly never again have to worry about where the next meal was coming from, and would eventually be able to assist his family financially. Among other things, he would repay Theresia's generosity by providing handsomely for the education of her three sons – in spite of the lack of a dowry, she did marry – two of whom became medical doctors (the third died as a young man).

Alongside his religious duties as a novice, Mendel came under the influence of Fr Matthaeus Klacel, a botanist who was already carrying out plant-breeding experiments at the monastery. Klacel was no dilettante: he was a member of the Prague Natural Science Society and the Brünn Agricultural Society, seriously interested in mineralogy and astronomy, and a respected philosopher who corresponded with academic philosophers at the universities. In other words, with just a little hyperbole, he was just an average member of the Augustinian community of St Thomas in Brünn. At one time, Klacel seemed likely to become Professor of Philosophy at the University of Prague, but lost the opportunity in the backlash against the revolutionary activity of 1848.

Mendel settled in well at the monastery, and attended a four-year course of theological study at Brünn Theological College. The monks were not directly affected by the upheavals of 1848, although Gregor must have been pleased about one consequence – the abolition of the practice of forced labour under which his peasant father had suffered.

Several of the older monks died while Mendel was still a student, and the community was so hard-pressed to make up the numbers to carry out its daily services that

Mendel's ordination was rushed through. It took place as early as was legally permitted, on the day he became 25, before he had finished his theological studies. He was ordained subdeacon on 22 July 1847, deacon on 4 August, and two days later, a fortnight after his 25th birthday, he became a full priest.

Mendel completed his theological course in June 1848, and was assigned duties as a parish priest and chaplain to the nearby hospital. The work did not suit him, and close proximity to the suffering of the sick made Mendel himself physically ill. We don't have the exact details of the illness, but Abbot Napp (who was in any case well aware of Mendel's academic bent) was sufficiently concerned that he found other work for him. This was easy, since Napp was by now Director of higher school education in Moravia, and in September 1849 he despatched Fr Mendel, as he now was, to work as a supply teacher at the *Gymnasium* in Znaim.

Mendel took to this like the proverbial duck to water. He taught mathematics and classics, and was popular both with the students and with the other members of staff at the school. The obvious next step was to formalize his position by getting a teaching certificate, which he could do simply by taking an examination set by the University of Vienna (some written papers, completed in Brünn, followed by an oral examination in Vienna itself), without attending any courses. He took the examination in 1850, and to the surprise of Mendel himself and his colleagues, he failed.

Mendel's written answers to the examination survive, together with the comments of his examiners. Although he did reasonably well in physics and mathematics, he failed – ironically, in view of his later achievements – in natural history. The examiners displayed no prejudice against this country priest: the papers show that Mendel really did have a very poor grasp of academic biology in 1850.

But that didn't make Mendel any less useful as a supply teacher in Brünn. His reputation was so high that, in spite of failing this examination, in the spring of 1851 he was asked to help out at the Technical College when one of the professors fell ill. He carried out the professor's teaching duties until early June, and was rewarded both with the appropriate pay for the job and with a testimonial from the Director,

> bearing witness to the fact that during your term of work here you have displayed great zeal, have given most useful instruction, have been most considerate to your pupils, and have manifested the most praiseworthy good feeling towards all the members of the School.

The examiners in Vienna had also been impressed by Mendel's enthusiasm and ability, and had only failed him because of the obvious gaps in his knowledge. Indeed, the chairman of the examiners had advised Mendel to take a university course, and pressed Mendel's case (probably at Gregor's request) with Abbot Napp during the summer of 1851. Napp needed little persuading. On 3 October 1851 he replied that

> The views you have been good enough to express regarding Father Gregor Mendel have decided me to send him to Vienna for a higher scientific training. I shall not

> grudge any expense requisite for the furtherance of this training . . .

And to his Bishop, Napp wrote to explain that:

> Father Gregor Mendel has proved unsuitable for work as a parish priest, but has on the other hand shown evidence of exceptional intellectual capacity and remarkable industry in the study of the natural sciences . . . for the full practical development of his powers in this respect it would seem necessary and desirable to send him to Vienna University where he will have full opportunities for study . . .

The Bishop approved,

> provided that in Vienna the above-named priest shall lead the life proper to a member of a religious order, and shall not become estranged from his profession.

Later that month, Gregor Johann Mendel arrived in Vienna to study physics at the university. He was now 29 years old, and Vienna was still one of the great capitals of Europe, with one of the great universities, where the Professor of Physics was Christian Doppler, of Doppler effect fame. And to put the Vienna of Mendel's day in a broader context, Johann Strauss the Younger was 26 the year Mendel arrived in Vienna, 16 years before Strauss would write 'The Blue Danube'.

Although Mendel did nothing to disgrace his holy orders while in Vienna, neither is there evidence that he led the life of a recluse. Surviving letters show that he enjoyed his time there, although he studied hard and had to spend the university holidays attending to religious duties back in Brünn. One highlight of his years as a university student was the marriage of his sister Theresia, which he attended in Heinzendorf in October 1852. This was his first opportunity to meet her husband, Leopold Schindler.

In Vienna, Mendel studied experimental physics, learning how to determine general laws from observation and experiment. He also studied statistics and probability, and learned the atomic theory of chemistry, planting firmly in his mind the idea that complex systems are made up of relatively simple fundamental units. He also took steps to fill the gaps in his education that the examiners had noted, attending courses in biological subjects, including plant physiology.

In an echo of the atomic theory of matter, the university was just beginning to teach that plants are made up of cells. It was also around this time that botanists first showed conclusively that plants reproduce sexually, and that contributions from both 'parents' were important in the development of a new plant from a seed. There was a growing interest among botanists in plant hybridization as a means to investigate how reproduction occurs. The books and papers left in Mendel's own library when he died (most with notes in his handwriting) show that he was well aware of these new developments in the early 1850s.

All of Mendel's studies in Vienna were crammed into two academic years, from October 1851 to the end of the summer term in 1853. The material covered in that short period shows that he cannot have had much time for fun and games, but in spite of his intensive study he did not actually take a degree. That was not the object of the exercise, and even without a degree he was now better qualified than ever to teach in Brünn while picking up the threads of his religious life and starting to carry out research of his own in his spare time.

There is no record of what Mendel did during his first year back at the monastery, but in May 1854 he was sent as a supply teacher to the Modern Technical High School, teaching physics and natural history to the lower classes and looking after the natural history collection. The Technical High School had been founded just two years before, and was one of the first schools in the region to provide a more technical (that is, scientific) education, instead of relying on the classics for its syllabus.

Mendel taught there for 14 years, and during that time the number of students never fell below 745 (the highest number during his time there was 986 pupils). Mendel's own position in the school remained supply teacher for all that time, even though he had been to university in Vienna, and even though he took the required teacher's examination again in 1856.

No records survive of those examinations.

All we know is that Mendel returned from Vienna after the oral in a depressed state, and that, since he remained a supply teacher, he cannot have passed. One possibility (consistent with a certain stubbornness in Mendel's character that will be highlighted later) is that he argued with the examiners about the theory of plant fertilization, and rather than withdraw his criticism of their ideas, he withdrew from the examination to avoid the indignity of being formally failed. This is supposition, but plausible, since one of the examiners, Professor Fenzl, was a die-hard believer in the old idea that a new plant develops only from the pollen cell, whereas Mendel's notes show that he had taken on board the new evidence for sexual reproduction in plants.

Whatever the reasons, it was in 1856 that Mendel began his intensive study of the way heredity works in peas, a painstaking and accurate series of experiments that lasted for seven years, and provided the evidence which reveals the fundamental laws at work in reproduction. Yet this was a strictly part-time activity, fitted in by Mendel in terms of both time and space. All the space he had to work in was a small strip of the monastery garden (a plot 35 metres long and 7 metres wide) where he could plant what he liked, and a greenhouse. All the time that he had was what could be spared from his full-time teaching post and his religious duties.

There is very little to say about Mendel's life outside his work in those years, because very little is known, and also because there simply cannot have been time for much life outside these activities. Mendel's first biographer, Hugo Iltis, had the opportunity to speak to some of his former pupils from the Technical High School, but that was more than 50 years after Mendel stopped being a teacher. By then, the memories of those pupils must have been coloured both by distance and by the fact that Mendel was already famous when the interviews were conducted. Nevertheless, Iltis's book *Life of Mendel* (Allen & Unwin) provides the only eye-witness picture we have of his time as a teacher.

We learn that Mendel was a man of medium height, already distinctly on the chubby side (as others have commented, after about 1860 the best instant image of Mendel is as a kind of jolly Friar Tuck). When carrying out his duties as schoolmaster, he usually wore the regular garb of a schoolmaster in those days of the Austrian Empire – 'tall hat; frock coat, usually rather too big for him; short trousers tucked into top-boots.' He was a jolly, friendly man, 'his blue eyes twinkling in the friendliest fashion through his gold-rimmed glasses,' with brown curly hair and a 'provincial burr' in his speech that amused his students. His down-to-earth nature showed when he was describing his plant-breeding experiments to his pupils, and used straightforward terms to describe their sexual process of reproduction. When the inevitable titters ran round the classroom, Mendel remonstrated with the boys, telling them how silly it was to be embarrassed about natural things.

Mendel loved all animals, and would confiscate catapults in an attempt to stop the slaughter of birds (but he had a horror of snakes, which he would not touch). He was so kind to his students that if one of them had a poor result in the end-of-term examinations, he would give him another set of (easier) questions to answer, so that his grade could be improved. Mendel's reasoning was that it was more important, at that stage of education, to encourage his students' enthusiasm for science than to stuff their heads with facts learned by rote – an astonishingly modern and utterly sensible view, which many present-day teachers could learn from. In return, the pupils loved him, and often visited the monastery in droves, 'like beetles buzzing at an open window'. There they would be shown the monastery garden, and told (without much understanding) about the experiments Mendel was carrying out. He seems to have been able to maintain class discipline effortlessly, though, by being so kind and considerate that nobody ever wanted to upset him.

In his work as a teacher, in spite of his indulgence towards the less able pupils, and in spite of his failure to obtain a teaching certificate, Mendel always received good reports. In the bureaucratic Austrian Empire, even the humblest schoolteacher had a file in the office of the provincial governor, and every year the file was added to with reports on the subject's suitability for his work. The phrases that recur in Mendel's file are 'complete satisfaction', 'zealous and successful endeavours', 'admirable', and the like.

There were two occasions on which sad news from the outside world penetrated Mendel's happy existence. His father died in 1857, and his mother in 1862. His letters show that Mendel was naturally distressed on both occasions, but these losses did not bear so heavily on a monk surrounded by his religious 'family' as they would on a solitary man.

Not that Mendel was cut off in Brünn. There is evidence, including a photograph, that later in 1862 he travelled to London. The details are far from clear, and it is hard to see how an Augustinian monk can have gone to England as a mere tourist. In a way, the timing was unfortunate, since although Darwin's *Origin* had been published in 1859, Mendel read the German translation of 1863, and may have been unaware of Darwin's work at the time he visited England. One thing we know for sure is that Darwin and Mendel never met: although both were in England at the same time, they were never in the same town at the same time. They would certainly have had plenty to talk about, though. This was just at the time when Mendel's long series of experiments with peas was nearing completion, and by 1862 he must have had a good idea of how the results were turning out.

As we have said, Mendel did not hit on peas for his study by chance. In 1856, he had been studying the reproduction of different kinds of plant for several years, was up to date with the latest theories about sexual reproduction (more up to date than at least one of the professors in Vienna!) and had also been breeding mice in his room at the monastery. In some ways, it might have been easier to work with mice – they breed more rapidly than peas, cramming many generations into a year. But Mendel never discussed any experiments he had carried out with the selective breeding of mice, very probably because he realized that such work with mammals would be seen by his superiors in the Church as unsuitably close to home for human beings created in God's image. Plants were safer, and Mendel soon discovered that peas were ideal, because they had several characteristics which could be easily identified in different generations, and which – crucially – lent themselves to statistical analysis.

You can get some idea of Mendel's

thoroughness from the fact that between 1854 and 1856, in a preliminary series of studies, he experimented with 34 varieties of edible pea before settling on the species used in his main study. He chose those which bred true and had seven easily identifiable characteristics, including the colour and shape of the pod, the position of the flowers, and the two we shall concentrate on here: the colour of the pea seeds themselves (either yellow or green) and the shape of the seeds (either smoothly round, or rough).

Mendel's training as a physicist shone through in the way he carried out the experiments. He wanted large numbers of results, so that he could draw statistically meaningful conclusions, and worked with a total of some 28,000 plants, of which 12,835 were 'carefully examined' (hardly the work of a gardener with a passing interest in science!). Previous researchers had bred large numbers of plants, to be sure, but in what seems to modern eyes a careless and unsystematic fashion, letting the plants breed more or less as they liked, and then trying to extract some understanding of what was going on from the confusion of hybrids produced. Mendel treated each plant individually, and studied each generation separately, so that he could follow the inheritance of characteristics cleanly from one generation to the next, and (crucially) on into subsequent generations.

What he found can be expressed very simply and very clearly. But remember that each experiment took a year to complete, and think of those 28,000 plants, each individually identified by Mendel, and the 12,835 studied in detail. Remember also that in order to carry out these experiments, Mendel had to fertilize the plants by hand, brushing the pollen from one 'parent' on to the flowers of the other, and taking care not to let flowers get fertilized accidentally.

In one of the most striking experiments, Mendel took plants from a variety with smooth seeds and crossed them with plants from a variety with rough seeds. In the next generation of plants, all the seeds were smooth. At first sight, it looked as if roughness had disappeared, or been bred out. But parents from this next generation produced in their turn plants in which 75 per cent of the seeds were smooth and 25 per cent rough (the actual figures were 5,474 smooth seeds and 1,850 rough seeds). Roughness had returned in a significant proportion of the 'grandchildren' of the original plants.

Mendel carried out further experiments in which all plants in this grandchild generation were bred to make further generations for analysis. We shan't go in to all the details, but shall move on to the conclusions that Mendel drew from these and hundreds of other experiments. And we shall translate the results into the language of biology in the twentieth century.

It was clear from Mendel's experiments that there is something in a pea plant that determines the properties of its seeds. That something we now call a *gene*, and the property it determines (in this case, seed shape) we call a *characteristic* (the overall set of characteristics making up the living form is called a *phenotype*). Each plant carries two copies of the gene (strictly speaking, different versions of the same gene are called *alleles*, but we shall stick to 'gene' here). In the original generation of this experiment, all the smooth-seed plants carried two copies of the same gene (SS), which specified the characteristic smoothness (we now know that each cell of the plant carries two copies of each gene). Similarly, each of the plants with rough seeds had two copies of the gene for roughness (RR).

In the next generation, each plant inherits one version of the gene from each parent. So each plant has one gene for roughness, and one for smoothness (RS). But the seeds are not each half smooth and half rough. The roundness gene is the *dominant* gene (a term introduced in this context by Mendel himself), and is the only one that is *expressed*. The roughness gene is said to be *recessive* (again, Mendel's terminology).

In a grandchild generation, though, there is a greater variety of possibilities. A plant may inherit either of the two genes from either parent. The possible combinations are RR, RS, SR and SS. So it is very easy to see that one-quarter of the plants will inherit the smooth gene from each parent, and one-quarter will inherit the rough gene from each parent. These offspring will have smooth and rough seeds, respectively. The rest will inherit the smooth gene from one parent and the rough gene from the other (25 per cent get the smooth gene from the 'father' and the rough gene from the 'mother', 25 per cent the other way around). Once again, where there is a choice only the dominant smooth gene is expressed in the characteristic. So out of all the plants in the third generation, only the 25 per cent of seeds that inherit the roughness gene from both parents will actually be rough.

There have been some modern critics of Mendel's work who thought that his published numbers were too close to the precise statistical values predicted by this kind of theory to represent the actual results of his experiments. It has been suggested that his numbers were less precise, and that he massaged the figures to get closer approximations to precise ratios, such as 3 : 1. But it turns out that those statistical critics were not very well versed in botany, and had forgotten to allow for the fact that, as any gardener knows, about 10 per cent of the seeds you plant fail to germinate. The minor discrepancies in the way Mendel counted plants and seeds from one generation to the next are, in fact, all accounted for by this natural wastage.

Mendel had shown conclusively that during sexual reproduction inheritance works not by blending characteristics from the two parents, but by taking individual characteristics from each of them. The reason why this is not obvious in, say, human beings is that there are so many genes expressing so many characteristics that the overall body you end up with is made up of thousands of characteristics inherited from your mother and thousands more (roughly the same number) inherited from your father. They all blur together to give an impression of blending inheritance – rather like the way the thousands of dots in a pointillist painting blur together when viewed from a distance to give the impression of a blend of colour.

Mendel's work was so precise and painstaking that he could study the effect of two genes simultaneously, crossing pea plants whose seeds were smooth and yellow, for example, with plants whose seeds were rough and green. He showed that each characteristic is passed on separately (we now know, by its own gene) and that the inheritance of yellowness, say, is independent of the inheritance of smoothness.

Even in Mendel's day, the natural thing for a university scientist to have done next would be to publish his results, perhaps in a paper in a learned journal, or perhaps as a scientific monograph. But this is where Mendel's relative isolation from the mainstream of scientific and academic life did hinder the dissemination of his results. Mendel actually presented his key discoveries about inheritance in plants (including the results discussed here, but much more besides) in two papers that he read to the Natural Science Society in Brünn in 1865, when he was 42 years old.

The Society had been established at the end of 1861, in the less restrictive atmosphere which followed the political changes of 1859, when the autocratic system of government was replaced by something marginally more liberal. It was a highly respectable institution, including in its ranks well-regarded scientists from many disciplines, particularly the life sciences. It has to be said, though, that it was a provincial organization, not in the same league as the great scientific societies of the time such as the Royal Society in London or the French Academy. The proceedings of its meetings were not regarded, as we shall see, as essential reading by scientists in other countries (or even in Vienna).

The Brünn Natural Science Society

exchanged copies of its published proceedings with 120 other learned societies, including those in Vienna, Berlin, St Petersburg, Uppsala, London and Rome. As a matter of course, Mendel's two lectures to the society were printed in the proceedings and circulated in this way. In addition, Mendel himself had reprints of the published version (which, he told correspondents, were exactly as the lectures were given), forming a monograph which he sent to a few correspondents.

Popular accounts of Mendel's life often gleefully point out that when his work was rediscovered in the early 1900s, and people went to look up the original papers, in some of the copies of the proceedings deposited in the libraries of those learned societies, the relevant pages had not even been cut, proving that nobody had ever read them. (Books were often issued with their pages uncut: the large sheets on which several pages were printed were folded and bound, and the purchaser was left to slit the edges of joined pages with a paper-knife.) The story is true, but it is equally true that in other copies the relevant pages had been cut, and the work had been read and noted by more than a few scientists across Europe.

The most intriguing thread of communication about Mendel's work that can be established starts with the German botanist H. Hoffmann, who mentioned Mendel's work in a book published in 1869. Another German botanist, W.O. Focke, picked up the work from Hoffmann's book, and mentioned it in a publication of his own in 1881. Charles Darwin owned a copy of Focke's book, but seems not to have noticed the mention of Mendel's work (the book was published less than a year before Darwin died). Shortly before he died, though, Darwin loaned his copy of Focke's book to George Romanes, an acolyte who (together with Darwin's son Francis) was a great support and help to Darwin in his final years. It was Romanes who wrote the entry in the *Encyclopaedia Britannica* which mentioned Mendel's work. Sadly, Romanes himself died in 1894, at the age of 46, and never lived to see the fusion of Darwinian and Mendelian ideas that formed the basis of the modern theory of evolution by natural selection.

There were also references to Mendel's work in other publications that appeared in Germany, Sweden, Russia, England and the

United States between the 1860s and the end of the nineteenth century. But the biggest surprise is that it wasn't promoted and brought to the attention of the wider community of biologists by Karl Wilhelm von Nägeli. Nägeli had been born in Switzerland; by the time Mendel's results were published by the Natural Science Society in Brünn, he was Professor of Botany at the University of Munich. He was a leading investigator of plant reproduction and hybridization, and had made a special study of the hawkweed family. He had doubts about Darwin's theory of natural selection, and leaned more towards the evolutionary scheme of Jean-Baptiste Lamarck – based on the idea that characteristics acquired in one generation could be passed on to the next.

Still, he was a natural person for Mendel to make contact with, and as Mendel's own plant-breeding experiments continued in the 1860s (including experiments with hawkweed) he began a correspondence with Nägeli. He sent a copy of the monograph to establish his credentials (the covering letter was written on New Year's Eve, 1866), and discussed his results in a series of ten letters that together make up a body of work that any modern scientist would consider well worth publishing. Although Mendel did present a paper on his hawkweed studies to the Brünn Natural Science Society in 1870 (and it was published in the proceedings in the usual way), it did not develop the ideas outlined in his pea-breeding papers, and the letters are much more informative.

At first, Nägeli was rather cool to this obscure monk with his new ideas about reproduction. He failed totally to accept the idea that characteristics present in one generation could be 'lost' in succeeding generations, as would happen, say, when RS peas are crossed with RR peas to produce RR varieties. He believed (but with no evidence to back himself up) that repeated breeding from the resulting strains of rough peas would at some stage yield smooth varieties again. But although he never wavered in this view, as the correspondence developed he began to change his tone. And in a letter dated 27 April 1870 and referring to some of Mendel's later experiments, he went so far as to say 'I consider myself indeed fortunate to have found in you such a skilled and successful colleague.'

The work that changed Nägeli's mind is described in the ten letters from Mendel, dealing with his plant-breeding experiments with other species, after the work with peas was completed. These letters form the best surviving record of Mendel's later work, confirming (if confirmation were needed) that the study of peas was no fluke, but part of a carefully planned and expertly carried out long-term programme of research. In a letter to Nägeli (the last) written on 18 November 1873, Mendel used the expression 'struggle for existence' in its Darwinian context, indicating his acceptance of the Darwinian theory. This may be one reason why Nägeli didn't promote Mendel's work more vigorously.

But this was virtually the end of the correspondence – not through any reluctance on Nägeli's part, but because Mendel became preoccupied with other matters. A recurring theme in the correspondence is Mendel's regret that nobody else – including Nägeli – had made the effort to reproduce his experiments and see the results for themselves. This brings home the point that where Mendel was ahead of his time was not so much in his ideas, for evolution and reproduction were hot topics among biologists in the second half of the nineteenth century. It was in his methods, the painstakingly accurate methods of experimental physics, taken over into botany.

Nägeli wrote again to Mendel in both 1874 and 1875, but received no reply. By then, to all intents and purposes Mendel's scientific career was long since over, as he had devoted himself to administration – the classic fate for a scientist, although in his case he was administering a monastery, not a university department.

Although Nägeli did not refer specifically to Mendel's work in his own publications, in a major book published in 1884 (the year Mendel died) he discussed reproduction in terms which in places echo Mendel's ideas, suggesting that he had not forgotten them entirely. And a year later, in another book by Nägeli, Mendel's monograph was listed in the bibliography, although not referred to specifically in the text. This, it seems, was very much Nägeli's style. He seldom referred to other scientists by name in his writings, and in the book of 1884, for example, there is no mention of the name 'Lamarck', even though

Nägeli's evolutionary ideas owed so much to Lamarck. So an opportunity to spread the word about Mendel's discoveries was turned down.

We said that Mendel stopped being a scientist when he was elected abbot, but that isn't strictly true. Although he no longer had time for the kind of research needed to continue his studies of reproduction, he now had all the space he could want in the monastery garden, and used this to good effect, becoming a well-known horticulturalist. He developed improved strains of apples, pears and vines, and took particular delight in developing varieties of fuchsia (one of which he named 'evolution'). Even the pea-breeding experiments were turned to practical use. When one of the varieties he developed turned out to be particularly tasty, it soon became a regular feature of the refectory table at the monastery of St Thomas. The official monastery gardener, Josef Maresch, worked under Mendel's direction, and described Mendel in glowing terms after his former abbot had become famous – 'what a gardener our prelate was! . . . there is not a single gardener who could not have learned from him!'

In the world outside the monastery, Mendel paid for a programme of replanting on a denuded mountainside. He chose the species to be planted, and took particular care to ensure that there were many varieties of flowers for bees to take nectar from – among his many activities, Mendel was an expert beekeeper. As well as producing delicious honey, Mendel's bees were, we need hardly add, the subject of some (unsuccessful) breeding experiments. He became Vice-President of the Brünn Apicultural Society, and was offered the Presidency in 1871, but declined. In his capacity as Vice-President (but at his own expense), Mendel travelled to Kiel in 1871 to attend the Congress of the League of German Beekeepers.

Mendel was also fascinated by the weather, and became an expert meteorologist. There was a keen interest in meteorology in Moravia, and a Meteorological Society had been established in Brünn in 1816. It eventually became part of the Natural Science Society, and from 1856 Mendel was in charge of processing its weather observations. He also made his own observations, right up until a few days before his death, recording them in his meticulous, neat handwriting. In 1857, in a fine early example of improving the public understanding of science, a list of seventeen experts was drawn up who could answer questions from the public in the Brünn area on scientific topics; the topic next to Mendel's name was meteorology.

Mendel became a founder-member of the Austrian Meteorological Society in 1865, and corresponded with meteorologists in other countries. The records from Brünn were of considerable importance in the development of an understanding of European weather patterns, and in 1881, for example, the eminent Dutch meteorologist Christoph Buys-Ballot wrote to Mendel from Utrecht soliciting updated information to assist in his work on European weather. To the public in Brünn, though, Mendel's fame as a meteorologist was sealed when he gave a talk to the Natural Science Society in November 1870, discussing a freak whirlwind that had struck the town on 13 October that year. His account (again, preserved in the Society's proceedings) showed great insight into the physical processes behind the creation of a whirlwind, as well as a dramatic literary style.

He described the scene of destruction in vivid detail, and must have impressed his audience almost as much as the whirlwind itself had.

Mendel also became fascinated by sunspots, obtaining a telescope which he used to make meticulous observations of the way these dark blotches appear on the surface of the Sun. (Of course, like all astronomers he projected the image of the spotty Sun onto a white surface, for looking through a telescope at the Sun can cause instant blindness.)

Like so many people down the years, Mendel became convinced that the changing spottiness of the Sun influences the weather in some way, but (also like many people down the years) he never found a reliable way to use sunspots in weather forecasting. He noticed the relationship between solar activity and the auroras. From November 1882 there is reference in his notebook to the appearance of a large group of sunspots at the same time as a large auroral display visible from Geneva and Poland, while telephone communications were disrupted across the United States. But by then all of these interests were incidental to Mendel's time-consuming work as abbot

– although, it has to be admitted, some of the most unpleasant aspects of that work were largely a result of Mendel's own stubbornness.

Cyrill Franz Napp died early in 1868, and the Augustinians at the monastery of St Thomas in Brünn had to elect a new abbot from among their members. There were only thirteen of them, and anyone over the age of 50 could pretty much be discounted because of the death duty problem that we mentioned earlier. Equally, anyone under the age of 40 would be too young to have the experience needed to fill the post. In the event, only twelve monks voted, because one of them was seriously ill. The election took place at the end of March 1868, when Mendel was in his 46th year – an ideal age from the point of view of the community. He had a high standing as a respected member of the local community, well known through his teaching; he was also respected as a member of learned scientific societies, experienced in dealing with the outside world, and must have been seen by his brothers as following very much in the mould of Abbot Napp.

Mendel must have known that he had a

good chance of being elected, although in a letter to his brother-in-law Leopold Schindler (dated 26 March 1868) he tried not to let his hopes run away with him:

> It is quite uncertain which of us will be the lucky one. Should the choice fall on me, which I hardly venture to hope, I shall send you a wire on Monday afternoon. If you don't get a telegram, you will know that someone else has been elected.

The Schindlers did get the telegram.

There is no doubt that Mendel did want the job. One of the main reasons was that the post carried with it a good income, and he would at last be in a position to repay his sister properly for the financial help she had given him when he was a student. He was already providing practical help: through his connections, Mendel's nephews went to school in Brünn, where the eldest, Joseph, taught for a while before his early death. All three of the boys lived near their uncle's residence in Brünn during their time there, and spent Sunday afternoons with him, often playing chess with the abbot. In his new capacity, he would be able to provide for them right through their university studies, and get them settled in the world. But it meant giving up his own teaching post, and although at first he fooled himself into thinking that this would leave him with more time for scientific research, it was not to be.

The monastery was a wealthy and important institution in Moravian life, and as its prelate Mendel was automatically a member of the Provincial Assembly. He took up this post at a time of political change. In 1867, the autocratic Habsburg Emperor had been forced to cede power to parliament, and become a constitutional monarch. The major political party behind this change was the liberal Constitutional Party, which introduced many reforms, reducing the power not just of the Emperor but also of the Church (for example, in education) and passing it over to the State. They were opposed by the conservatives, who were dominated by the big landowners and by high-ranking Church representatives. In a further complication, the liberal reformers were largely associated with the German-speaking part of the Empire.

To the surprise of his colleagues in the

Church – and the annoyance of the Czech majority in Moravia – Mendel (as you might have guessed) supported and voted with the Constitutional Party. He thus incurred the displeasure of the bishop in Brünn, and split the small community of monks, who were roughly equally divided between Czechs and Germans.

Mendel also received other time-consuming administrative appointments, to add to his responsibility for running the large monastery estates (which included several dairy farms): Vice-President of the Natural Science Society, Vice-President of the Agricultural Society, Deputy Chairman and then Chairman of the Moravian Mortgage Bank (a real job, not a sinecure, which involved several hours of work several days a week), and more besides. In 1870, he was appointed to serve on a government committee carrying out the important task of adjusting the land tax in Moravia 'in reliance upon your well-tried loyal and patriotic sentiments and your thorough knowledge of agriculture and land values both in general and in particular'.

It is hardly surprising that in 1872 the Emperor conferred on Mendel the Order of

Franz Josef, 'in recognition of your meritorious and patriotic activities'. The award may have been only his due as prelate of an important monastery, but if it went with the job, he certainly earned it. Mendel travelled locally, overseeing the estates, and more widely: to Rome to pay his respects to the Pope (bringing back grape seeds that he planted, and from which he grew thriving vines in the monastery garden); to Berlin, to Vienna, to the Alps; a holiday in the Rhineland and another in Venice. The point of this rather breathless run through Mendel's activities during his years as abbot is simply to emphasize that he was no hermit-like religious recluse, but very much a man of the world.

For all that, the last ten years of Mendel's life were blighted by a stand that he took as a matter of principle, refusing to bend in a worldly-wise fashion. It all began in 1874, when the liberal government that Mendel himself supported (the irony was not lost on his political opponents) introduced a tax on monastic property, designed to offset government expenditure on religious affairs and, especially, to help provide for the stipends of parish priests. The tax was specifically to provide a religious fund, not to make a contribution to the general finances of the government. But Mendel decided that it was unconstitutional.

Like every other religious institution in the country, the monastery of St Thomas in Brünn was assessed. Its abbot was informed that its contribution to the fund would be 7,330 guilders per year for the five-year period 1875–1880. Gregor Mendel was the only abbot in the whole of the Empire who refused to accept the legality of the demand. He didn't even dignify the demand by going through the appeals procedure that had been set up to address situations where the institutions being taxed felt the assessment to be too high. By appealing, Mendel would have accepted the legality of the demand. Instead, he tried to fight the demand on legal grounds, going against the advice of the monastery's own lawyers.

He did offer to pay 2,000 guilders as a 'voluntary contribution' to the religious fund, this being, he told the authorities, the largest amount the monastery could afford. But since he refused to acknowledge the right of the authorities to tax his monastery, the offer was declined. Instead, in 1876 the government authorities sequestered part of the monastery's estates in lieu of payment. Still protesting that the law was unconstitutional, Mendel made it clear that even though the debt was paid there was no way he would pay the tax next year, or any year. So a large chunk of the monastery's property remained in the hands of the sequestrators while the argument rumbled on.

Mendel managed to embarrass and/or annoy just about everybody as the years went by. The Constitutional Party was annoyed that one of their own supporters should rock the boat; the conservatives were annoyed that he was defying the rule of government, even if it wasn't their government; the Church thought he was being foolish; his own monks were alarmed at the prospect of seeing the prosperous monastery ruined. Several times, the authorities tried to persuade Mendel that all he had to do was agree to the principle of paying the tax, and lodge an appeal, and they would settle a smaller tax burden on him. But he refused to budge.

Of course, you can't beat the government. The old saying is that the two certainties in life are death and taxes, and both duly came along. When Mendel died in 1884, his successor agreed to pay the back taxes, and the sequestered property was returned to the monks. But that isn't the end of the story. After repeated appeals, the monastery had its entire tax position re-assessed. Because Mendel had never been willing to make a full return of the income and expenditure of the monastery, the figures had always been based on estimates. Now, with a little judicious book-keeping sleight of hand, the monks were able to show that, after allowing for the expenditure of the establishment on salaries, and removing the value of the library and art treasures from the calculation, the monastery was actually running at a loss! So they should not have paid any tax at all.

According to the new figures, the monastery was due a refund of 19,876 guilders (allowing for the administrative costs of the sequestrators). After some humming and hawing by the government, they got the money back in July 1886, and were given exemption from paying the tax for the decade 1891–1900. In fact, they never did pay it. Like all the other Moravian monasteries, they ended up agreeing to pay the tax in principle, but finding that they had no income to tax in practice – a classic political compromise wholly alien to Mendel's straightforward way of dealing with the world.

Not that Mendel let the dispute ruin his life. He had the company of his younger nephews, Ferdinand and Alois, for much of this period (Ferdinand, the youngest of the Schindler boys, only left school in Brünn in 1883), his interests in horticulture, beekeeping and meteorology, and visitors – often from the world of science – who stayed as guests at the monastery. And he was famously generous to charity, which gave him real pleasure. Among other things, he contributed 3,000 guilders towards the construction of a fire station in Heinzendorf.

Over the last few years his health went into a gradual decline. He was overweight, had heart trouble, smoked small cigars in large numbers (as many as twenty a day), and suffered with kidney problems. He died in his sleep, at 2 a.m. on the night of 6 January 1884, two days after making his last meteorological observations. One obituary notice recorded that:

In his activities as abbot he won the respect and honour of all by his free-handedness, affection and kindliness, so that he can justly be said to have had no personal enemies. He never denied help to any who applied for it. Prelate Mendel had the rare gift of being able to bestow alms without letting the petitioner feel any sense of dependence.

He summed up his own view of his life to one of his colleagues shortly before he died:

> Though I have suffered some bitter moments in my life, I must thankfully admit that most of it has been pleasant and good. My scientific work has brought me a great deal of satisfaction, and I am convinced that it will not be long before the whole world acknowledges it.

In fact, it would take just a quarter of a century for Mendel to achieve the scientific recognition he deserved.

Afterword

As early as 1889, five years after Mendel had died (and seven years after Darwin's death), the Dutch botanist Hugo de Vries published a book, *Intracellular Pangenesis*, in which he tried to explain Darwin's ideas about evolution in terms of the developing understanding of how cells worked. ('Pangenesis' was a term invented by Darwin, and eventually gave us the term 'gene'.) Over the next ten years or so, de Vries carried out many plant-breeding experiments similar to Mendel's, and reached similar conclusions. But it seems that de Vries was unaware of Mendel's work while he was carrying out these experiments, and only came across Mendel's papers in 1899, when he was preparing his own work for publication and was carrying out a careful literature search in order to put it into context.

He must have made the discovery with very mixed feelings. On the one hand, here was confirmation that he was working along the right lines; on the other, he had been preempted! Perhaps those mixed feelings are expressed in the way de Vries published his own results, in two separate papers, both published in March 1900. The first, written in French, was very short, and made no mention of Mendel. The second, itself running to only eight pages, was written in German and refers to Mendel's work, commenting that 'this important monograph is so rarely quoted that I myself did not become acquainted with it until I had concluded most of my experiments'. In his summing up, de Vries gave Mendel pride of place:

> From these and numerous other experiments I drew the conclusion that the law of segregation of hybrids as discovered by Mendel for peas finds very general application in the plant kingdom and that it has a basic significance for the study of the units of which the species character is composed.

By the middle of 1900, though, there was an increasingly powerful reason for everybody engaged in this kind of research to acknowledge Mendel's priority – by doing so, they avoided an unseemly squabble among themselves. For de Vries was not the only one at the end of the nineteenth century to repeat Mendel's discoveries. A German botanist,

Carl Correns, had also been carrying out plant-breeding experiments (even working with peas for some of his investigations), and was getting his results ready for publication when he received a copy of de Vries' French paper. Just a month after de Vries, in April 1900, Correns published his own first paper on the subject. At the time he prepared it, he had seen only the French version of de Vries' work. He wrote:

> In my hybridization experiments upon various races of maize and pea, I have come to the same result as de Vries . . . I believed myself, as de Vries obviously believes himself, to be an innovator. Subsequently, however, I found that in Brünn during the 'sixties Abbot Gregor Mendel, devoting many years to the most extensive experiments on peas, had not only obtained the same results as de Vries and myself, but had actually given the very same explanation.

It didn't end there. An Austrian, Erich Tschermak von Seysenegg, was also carrying out the same kind of experiments at about the same time, and published his results in

June 1900. Even more than the other two, he spelled out the significance of Mendel's work, and took on board some of Mendel's terminology, such as 'dominance'. He wrote that:

> The simultaneous discovery of Mendel by Correns, de Vries and myself seems to me peculiarly gratifying. I, too, as late as the second year of my experiments, believed that I had happened upon something entirely new.

What was so gratifying about the rediscovery of Mendel was that it meant that none of the rediscoverers could claim credit. Disappointed though each of the various botanists might be that he could not claim all the glory for himself, at least each of them knew that neither of the others could claim to have got there first.

Everyone was eager to acknowledge Mendel's priority, and the way his work had been rediscovered – independently three times in the same year, right at the beginning of a new century – helped to ensure that biologists around the world seized on the news and carried out further experiments of the same kind. And when those biologists (notably William Bateson, in England) looked at Mendel's original papers, many of them began to recognize the importance of his methods – the methods of experimental physics applied to biology, the results analysed in a rigorous statistical fashion.

A full understanding of genetics and heredity had to await the discovery of the double-helix structure of DNA, and the subsequent cracking of the genetic code. Appropriately, though, those developments came in large measure from physicists who had turned to biology, such as Francis Crick. Mendel would surely have approved, both of the people involved and of their methods.

A brief history of science

All science is either physics or stamp collecting.

Ernest Rutherford

c. 2000 BC	First phase of construction at Stonehenge, an early observatory.
430 BC	Democritus teaches that everything is made of atoms.
c. 330 BC	Aristotle teaches that the Universe is made of concentric spheres, centred on the Earth.
300 BC	Euclid gathers together and writes down the mathematical knowledge of his time.
265 BC	Archimedes discovers his principle of buoyancy while having a bath.
c. 235 BC	Eratosthenes of Cyrene calculates the size of the Earth with commendable accuracy.

AD 79	Pliny the Elder dies while studying an eruption of Mount Vesuvius.
400	The term 'chemistry' is used for the first time, by scholars in Alexandria.
c. 1020	Alhazen, the greatest scientist of the so-called Dark Ages, explains the workings of lenses and parabolic mirrors.
1054	Chinese astronomers observe a supernova; the remnant is visible today as the Crab Nebula.
1490	Leonardo da Vinci studies the capillary action of liquids.
1543	In his book *De revolutionibus*, Nicholas Copernicus places the Sun, not the Earth, at the centre of the Solar System. Andreas Vesalius studies human anatomy in a scientific way.
c. 1550	The reflecting telescope, and later the refracting telescope,

	pioneered by Leonard Digges.
1572	Tycho Brahe observes a supernova.
1580	Prospero Alpini realizes that plants come in two sexes.
1596	Botanical knowledge is summarized in John Gerrard's *Herbal*.
1608	Hans Lippershey's invention of a refracting telescope is the first for which there is firm evidence.
1609–19	Johannes Kepler publishes his laws of planetary motion.
1610	Galileo Galilei observes the moons of Jupiter through a telescope.
1628	William Harvey publishes his discovery of the circulation of the blood.
1643	Mercury barometer invented by Evangelista Torricelli.
1656	Christiaan Huygens correctly identifies the rings of Saturn, and invents the pendulum

	clock.
1662	The law relating the pressure and volume of a gas discovered by Robert Boyle, and named after him.
1665	Robert Hooke describes living cells.
1668	A functional reflecting telescope is made by Isaac Newton, unaware of Digges's earlier work.
1673	Antony van Leeuwenhoeck reports his discoveries with the microscope to the Royal Society.
1675	Ole Roemer measures the speed of light by timing eclipses of the moons of Jupiter.
1683	Van Leeuwenhoeck observes bacteria.
1687	Publication of Newton's

	Principia, which includes his law of gravitation.
1705	Edmond Halley publishes his prediction of the return of the comet that now bears his name.
1737	Carl Linnaeus publishes his classification of plants.
1749	Georges Louis Leclerc, Comte de Buffon, defines a species in the modern sense.
1758	Halley's Comet returns, as predicted.
1760	John Michell explains earthquakes.
1772	Carl Scheele discovers oxygen; Joseph Priestley independently discovers it two years later.
1773	Pierre de Laplace begins his work on refining planetary orbits. When asked by Napoleon why there was no mention of God in his scheme, Laplace replied, 'I

	have no need of that hypothesis.'
1783	John Michell is the first person to suggest the existence of 'dark stars' – now known as black holes.
1789	Antoine Lavoisier publishes a table of thirty-one chemical elements.
1796	Edward Jenner carries out the first inoculation, against smallpox.
1798	Henry Cavendish determines the mass of the Earth.
1802	Thomas Young publishes his first paper on the wave theory of light. Jean-Baptiste Lamarck invents the term 'biology'.
1803	John Dalton proposes the atomic theory of matter.
1807	Humphry Davy discovers sodium and potassium, and goes on to find several other elements.

1811	Amedeo Avogadro proposes the law that gases contain equal numbers of molecules under the same conditions.
1816	Augustin Fresnel develops his version of the wave theory of light.
1826	First photograph from nature obtained by Nicéphore Niépce.
1828	Friedrich Wöhler synthesizes an organic compound (urea) from inorganic ingredients.
1830	Publication of the first volume of Charles Lyell's *Principles of Geology*.
1831	Michael Faraday and Joseph Henry discover electromagnetic induction. Charles Darwin sets sail on the *Beagle*.
1837	Louis Agassiz coins the term 'ice age' (*die Eiszeit*).
1842	Christian Doppler describes the effect that now bears his

	name.
1849	Hippolyte Fizeau measures the speed of light to within 5 per cent of the modern value.
1851	Jean Foucault uses his eponymous pendulum to demonstrate the rotation of the Earth.
1857	Publication of Darwin's *Origin of Species.* Coincidentally, Gregor Mendel begins his experiments with pea breeding.
1864	James Clerk Maxwell formulates equations describing all electric and magnetic phenomena, and shows that light is an electromagnetic wave.
1868	Jules Janssen and Norman Lockyer identify helium from its lines in the Sun's spectrum.
1871	Dmitri Mendeleyev predicts

	that 'new' elements will be found to fit the gaps in his periodic table.
1887	Experiment carried out by Albert Michelson and Edward Morley finds no evidence for the existence of an 'aether'.
1895	X-rays discovered by Wilhelm Röntgen. Sigmund Freud begins to develop psychoanalysis.
1896	Antoine Becquerel discovers radioactivity.
1897	Electron identified by J. J. Thomson.
1898	Marie and Pierre Curie discover radium.
1900	Max Planck explains how electromagnetic radiation is absorbed and emitted as quanta. Various biologists rediscover Medel's principles of genetics and heredity.
1903	First powered and controlled flight in an aircraft heavier

	than air, by Orville Wright.
1905	Einstein's special theory of relativity published.
1908	Hermann Minkowski shows that the special theory of relativity can be elegantly explained in geometrical terms if time is the fourth dimension.
1909	First use of the word 'gene', by Wilhelm Johannsen.
1912	Discovery of cosmic rays by Victor Hess. Alfred Wegener proposes the idea of continental drift, which led in the 1960s to the theory of plate tectonics.
1913	Discovery of the ozone layer by Charles Fabry.
1914	Ernest Rutherford discovers the proton, a name he coins in 1919.
1915	Einstein presents his general theory of relativity to the Prussian Academy of

	Sciences.
1916	Karl Schwarzschild shows that the general theory of relativity predicts the existence of what are now called black holes.
1919	Arthur Eddington and others observe the bending of starlight during a total eclipse of the Sun, and so confirm the accuracy of the general theory of relativity. Rutherford splits the atom.
1923	Louis de Broglie suggests that electrons can behave as waves.
1926	Enrico Fermi and Paul Dirac discover the statistical rules which govern the behaviour of quantum particles such as electrons.
1927	Werner Heisenberg develops the uncertainty principle.
1928	Alexander Fleming discovers penicillin.

1929	Edwin Hubble discovers that the Universe is expanding.
1930s	Linus Pauling explains chemistry in terms of quantum physics.
1932	Neutron discovered by James Chadwick.
1937	Grote Reber builds the first radio telescope.
1942	First controlled nuclear reaction achieved by Enrico Fermi and others.
1940s	George Gamow, Ralph Alpher and Robert Herman develop the Big Bang theory of the origin of the Universe.
1948	Richard Feynman extends quantum theory by developing quantum electrodynamics.
1951	Francis Crick and James Watson work out the helix structure of DNA, using X-ray results obtained by Rosalind Franklin.

1957	Fred Hoyle, together with William Fowler and Geoffrey and Margaret Burbidge, explains how elements are synthesized inside stars. The laser is devised by Gordon Gould. Launch of first artificial satellite, *Sputnik 1*.
1960	Jacques Monod and Francis Jacob identify messenger RNA.
1961	First part of the genetic code cracked by Marshall Nirenberg.
1963	Discovery of quasars by Maarten Schmidt.
1964	W.D. Hamilton explains altruism in terms of what is now called sociobiology.
1965	Arno Penzias and Robert Wilson discover the cosmic background radiation left over from the Big Bang.
1967	Discovery of the first pulsar

	by Jocelyn Bell.
1979	Alan Guth starts to develop the inflationary model of the very early Universe.
1988	Scientists at Caltech discover that there is nothing in the laws of physics that forbids time travel.
1995	Top quark identified.
1996	Tentative identification of evidence of primitive life in a meteorite believed to have originated on Mars.